Miki Marble™

Another Hare-Brain Science Tale

By John A. Honeycutt

Illustrations by Ana Nastevska

Art Direction by Kristina Ilievska

Production by Layne Petersen

With love to:
Jennifer, Danielle, Heather, Sara, and Cody

Miki Marble had twenty-nine marbles.

Miki Marble was afraid
of the dark, but had
a flashlight. Miki Marble
sometimes turned on the
flashlight at night.

Miki Marble had twenty-one friends. Miki Marble liked to fly kites, and had a long kite string made from cotton.

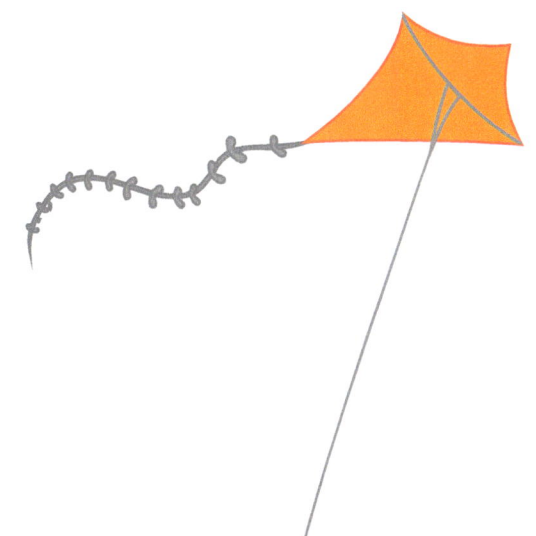

Also, for some reason, Miki Marble had some postage stamps.

Miki Marble got tired of everyone saying "Miki Marble" all the time. Miki Marble wanted a different name. Miki Marble asked for suggestions from each of the twenty-one friends.

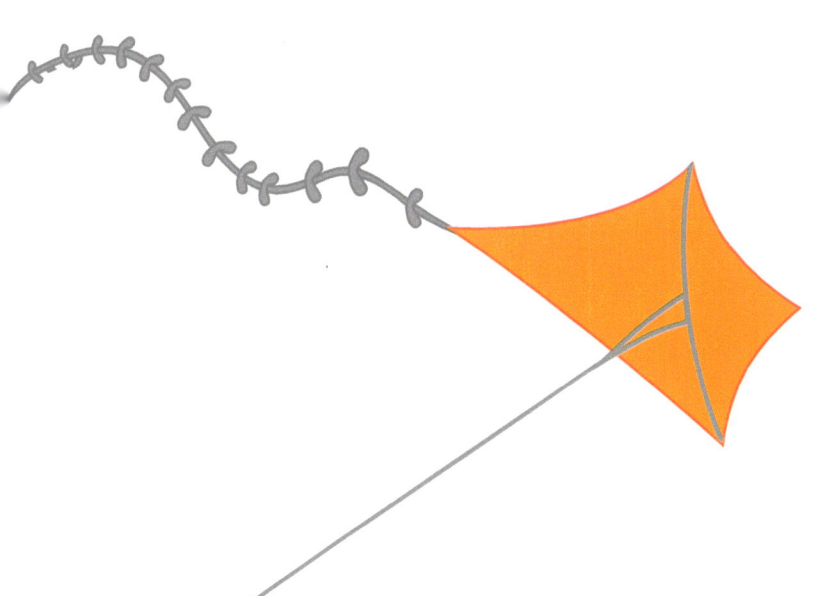

One of the friends said
"How about Miki Kite?"

Another said "How about Miki String?" Another said "How about Miki Stamp?"

Miki Marble didn't care
for any of these ideas.

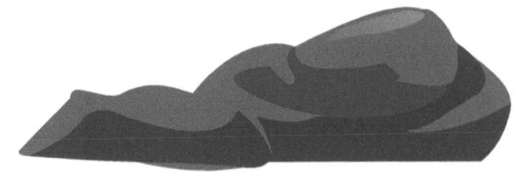

The next night,
Miki Marble had an idea.

Miki Marble took a rocket ship all the way to the sun, and then used the cotton string to measure distances.

Miki Marble measured the distance from the sun to each planet.

(Mercury)

The first planet was hot!

(Venus)

The next planet was
kind of like Earth.
This planet was kind of
cloudy.

(Earth)

The next planet
actually was Earth.

(Mars)

And the fourth
planet had a lot of
reddish rock.

Miki Marble noticed the first four planets were mostly made up of rock.

(Neptune)

(Uranus)

(Saturn)

(Jupiter)

The last four planets
were very different from
the first four.
They were big.
Really big.

Miki Marble measured these
distances with the cotton kite string.

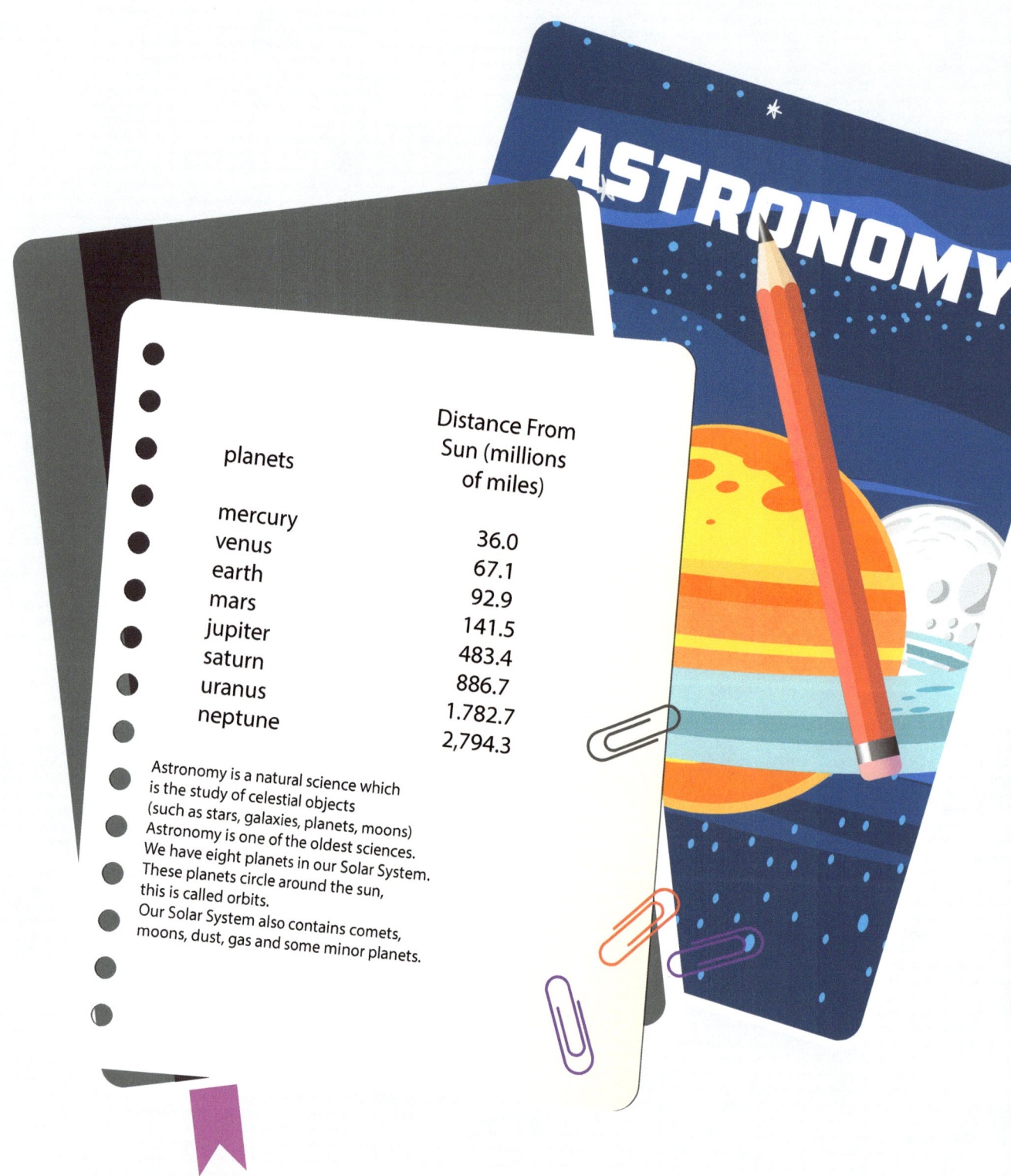

planets	Distance From Sun (millions of miles)
mercury	36.0
venus	67.1
earth	92.9
mars	141.5
jupiter	483.4
saturn	886.7
uranus	1.782.7
neptune	2,794.3

Astronomy is a natural science which is the study of celestial objects (such as stars, galaxies, planets, moons) Astronomy is one of the oldest sciences. We have eight planets in our Solar System. These planets circle around the sun, this is called orbits.
Our Solar System also contains comets, moons, dust, gas and some minor planets.

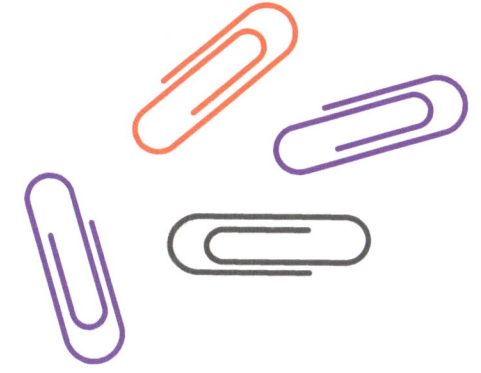

Miki Marble used the measurements to make a model for science class.

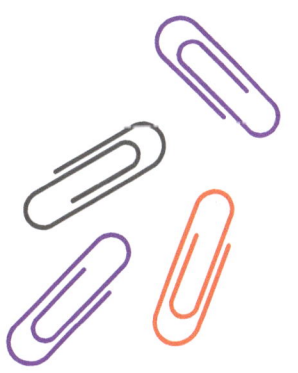

It was a solar system model. Miki Marble used marbles for planets. Miki Marble used the flashlight as the sun.

Miki Marble used kite string to show the distances.

From then on,
Miki Marble was known
as Piki Planet.

Piki Planet gave away the extra marbles to the twenty-one friends. Piki Planet placed a postage stamp on each marble in the model. Piki Planet wrote a description on each postage stamp.

Mercury	*hot and rocky*
Venus	*rocky and sort of like Earth*
Earth	*my home*
Mars	*really red*

Jupiter	*really huge and made out of gas*
Saturn	*huge, and with rings around it*
Uranus	*big and cold*
Neptune	*big and really, really cold*

MERCURY

HOT and rocky

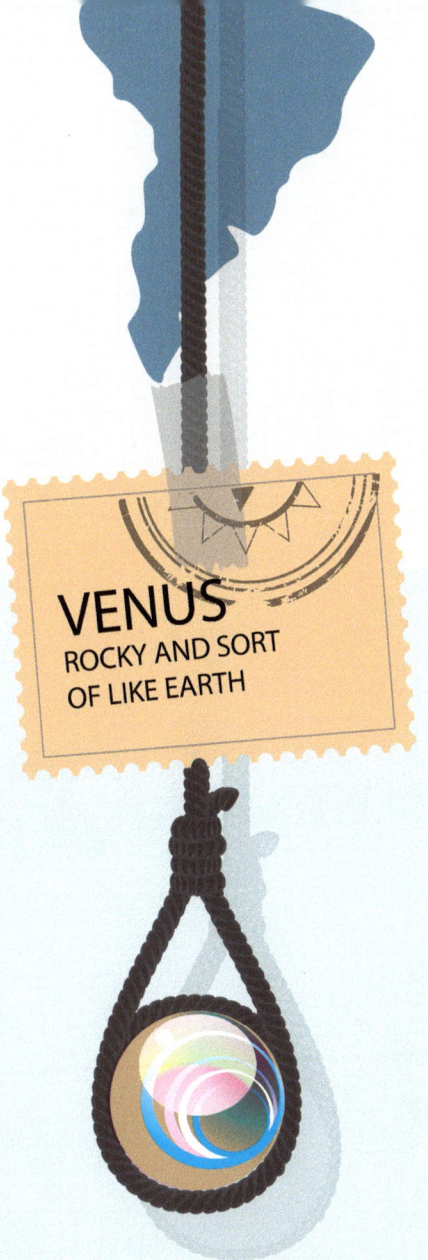

VENUS
ROCKY AND SORT
OF LIKE EARTH

EARTH
MY HOME

JUPITER
REALLY, REALLY
HUGE AND MADE OUT
OF GAS

SATURN
HUGE AND WITH
RINGS AROUND IT

URANUS
BIG AND COLD

NEPTUNE
big and really,
really cold

Piki Planet used his favorite marble for Mercury.

An opaque, white marble represented Venus.

Earth was a blue marble.

Mars was a red marble.

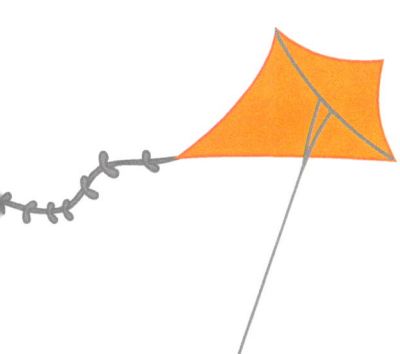

Extremely big marbles represented Jupiter and Saturn.

Two other big marbles represented Uranus and Neptune.

Both of these were placed in the freezer first – to make them very cold.

Piki Planet noticed something from the rocket ship trip.

The first four planets were similar. They were made up mostly of rock. (*Terrestrials*).

The last four were big! Like giant planets! (*Gas Giants*).

SCIENCE COMPETITION

Piki Planet entered the model into a science contest. The model did not win, but a lot of people looked at it and liked it.

Piki Planet's friends never used the name Miki Marble again. That made Piki Planet happy.

Oh, and something else - Piki Planet was never afraid of the dark.

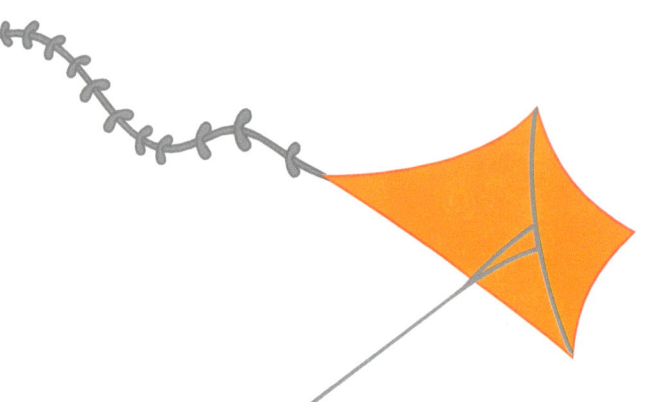

Oh, and just one more thing... Piki Planet used the kite for space travel after that — the rocket ship was fast, but not as much fun as the kite.

Planets and our Solar System

Our Solar System includes the Sun and the objects that orbit the Sun. The eight largest objects which orbit the Sun are called planets. Other objects that orbit the sun are smaller than the planets. The smaller objects include dwarf planets, comets and asteroids.

Scientists who study our Solar System – and outer space – are called Astronomists. Astronomy is a natural science. It is the study of celestial objects (such as stars, galaxies, planets, moons). Astronomy is one of the oldest sciences.

There are eight known planets in our Solar System. The planets circle around the Sun – they orbit the Sun. This is similar to how the moon circles around Earth.

Our Solar System also contains comets, moons, dust, gas and some minor planets.

For example, Pluto has been considered the ninth planet at times. As of the publishing of this book (2014) it is considered a dwarf planet, rather than a full-fledged planet. Pluto is farther from the Sun than Neptune. Also, compared to the eight planets described in this book, Pluto is much smaller.

Young scientists interested in Astronomy should realize that new discoveries about the Earth, the planets, the Sun, and our Solar System happen often. Astronomists keep learning new information. Astronomy is a very cool science job.

Learn more at www.Hare-Brain.com